Health 95

蟎虫家园

The Mite Farm

Gunter Pauli

[比]冈特·鲍利　著

[哥伦]凯瑟琳娜·巴赫　绘

何家振　译

上海远东出版社

目录

Contents

螨虫一家站在一座巨大的崭新大楼前面，他们正要去寻找一个好住处。

"我们进去看看地板上是否有地毯。"螨虫爸爸说。

"你知道，我不喜欢这些毛皮地毯。"螨虫妈妈说，"走在上面感觉很硬。"

A family of mites are standing in front of a grand, new building, looking for a good place to live.

"Let's go inside to see if it has carpets on the floors," says the dad.

"I don't like these furry floor tiles, you know," the mom replies. "They're hard to walk on."

我不喜欢这些毛皮地毯

I don't like these furry floor tiles

现在让我们看看窗户吧

Now let's have a look at the windows

"但是，这种地毯的好处也很多呢。我们一家人将拥有所需的全部食物。"

"是的，有很多人住在这里，他们的皮肤不停地脱落死细胞，为我们提供了食物。"

"现在让我们看看窗户吧。"螨虫爸爸说。

"But think of all the advantages. We will have all the food we need for our big family."

"That's true. There are a lot of people living here, and their skins shed dead cells all the time, providing us with food."

"Now let's have a look at the windows," says Dad Mite.

"你觉得他们会为了呼吸新鲜空气打开窗户吗？"

"不会的，在这些现代化的大楼里，他们从来不开窗户，他们为了节能不计代价，甚至不惜牺牲健康。"

"我们太幸运了，人们关上所有的窗户，强迫所有人在相同的空气中呼吸，天天如此。"螨虫妈妈说。

Do you think they'd want to open a window for fresh air?"

"Oh no, in these modern buildings, they never open windows. They want to save energy at all costs, even at the cost of their own health."

"We are so lucky that people seal all their windows and force everyone to breathe the same air, day in and day out," remarks Mother Mite.

……地毯是躲藏的好地方……

... carpets are good places to hide ...

"我不知道他们是怎么活下来的。但是我看到他们在空调系统中装上强大的过滤装置，滤掉他们不想要的东西。"

"这些地毯是我们躲藏的好地方。你知道，如果他们看见我们，就会设法杀死我们。"

"I don't know how they survive. But I see that they use strong filters in the air conditioning system to get rid of things they don't like."

"These carpets are good places for us to hide then. You know, if they see us, they will try to kill us."

"别担心，人们从来不看他们脚下有什么，特别是不看地毯里有什么。"

"人们允许我们住在这里，还允许我们把粪便留在这里，这真令人惊讶。他们把地毯弄得这么脏，有些人在这里根本无法呼吸。"

"Don't worry, people never look at what's below their feet, especially not at what's in the carpet."

"It's amazing how people would allow us to live here and leave our droppings behind. They make the carpets so dirty and make it hard for some people to breathe in here."

人们从来不看他们脚下有什么

People never look at what's below their feet

我们最坏的敌人——太阳

Our worst enemy - the sun

"住在这儿，我们世代都会受到保护，因为人们在窗户上贴上了防紫外线膜，以保护地毯和艺术品不会被太阳晒褪色。"

"我们真幸运！那就是说，他们保护我们不受最坏的敌人——太阳的照射，而且这里还有丰富的美食。"

"既然我们不能清除自己的粪便，那就得等着一年被迫洗一次化学浴了。"螨虫爸爸警告说，"那玩意儿毒性很大，就连人类接触它也会生病。"

"Well, we'd be protected for generations to come, because people put ultra-violet film on the windows to protect their carpets and artwork from being faded by the sun."

"Lucky us again! That means they protect us from our worst enemy – the sun – while we enjoy the abundant food available here."

"As we cannot clean up after ourselves, we have to be prepared to have a chemical bath once a year," warns Dad Mite. "And that stuff is so toxic that even people can get sick from it."

"等到他们弄来那些化学杀毒剂时，我们会变得更强大、更有抗药性。"螨虫妈妈说道。

"是的，暴露在这种化学喷雾剂中几年之后，我们就适应了，到时候它对我们根本不管用。我们懂得生存之道，绝不是好对付的！"

"By the time they come with that terrible onslaught of chemicals, we'd have become much stronger and more resistant," says Mother Mite.

"That's true. After being exposed to those chemical sprays for years, we've become used to it and it hardly bothers us anymore. We know how to survive and have become a force to be reckoned with!"

更强大、更有抗药性

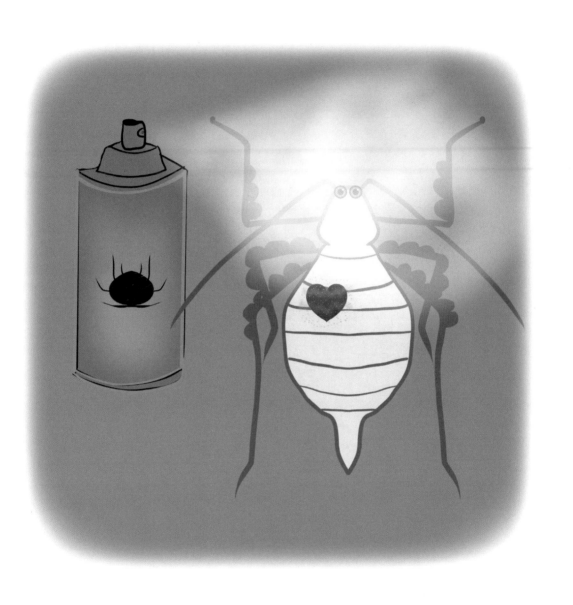

stronger and more resistance

我们的聚居地就越多

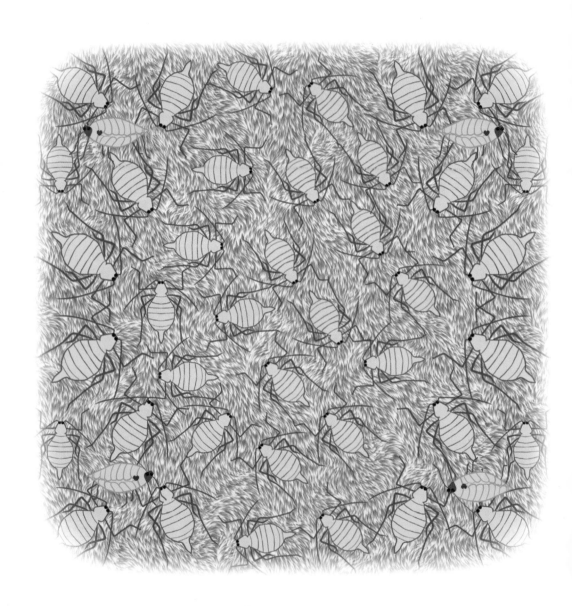

The more colonies we can establish

"是的，不管人们怎么做，比如，用可再生纤维做地毯，并且在本地制作以减少因交通运输产生的污染排放，我们都能轻易地生存，甚至更加兴旺!"

"地上铺的地毯越多，我们的聚居地就越多。"

"Yes, despite what people do, like making carpets from recycled fibres and doing so locally, to reduce emissions by not unnecessary transporting things around the world, we can easily survive – and even thrive!"

"The more carpets that are being laid, the more colonies we can establish."

"我知道。总部设在亚特兰大的全世界最大地毯制造商之一，不是曾经说过'可持续发展路途漫漫，我们只是迈出了很小的第一步'吗？"

"他说得对。不管他们用什么制造地毯，在这场游戏中的每一步，我们都注定会赢！"

……这仅仅是开始！……

"I know. Wasn't it one of the great carpet makers in the world, the one from Atlanta, who used to say that sustainability is only the small beginning of a long process?"

"And he was right. Whatever they make their carpets from, we'll certainly win at every step of this game!"

... AND IT HAS ONLY JUST BEGUN!...

······这仅仅是开始！······

... AND IT HAS ONLY JUST BEGUN! ...

Did You Know?

你知道吗？

地球上有48 000种螨虫。很多螨虫吃植物，有些啄食叶子，另外一些以鸟、鹿等动物为食，还有些寄生于蜜蜂等其他昆虫身上。

There are 48,000 different species of mites on Earth. Many eat plants, some tunnel through leaves, others feed on animals such as birds and deer, and some infest other insects like bees.

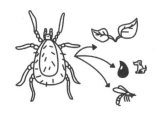

尘螨生长在房间里，特别喜欢待在温暖的地方，比如床上。它们以人皮肤的死细胞为食。另一些种类的螨虫以花粉和宠物毛屑为食。

Dust mites live in houses, especially in warm places like beds. They feed on the dead cells of human skin. Other types feed on pollen and pet dander.

尘螨只有半毫米大。尽管螨虫不咬、不叮、不打洞，但它们的粪便是过敏、哮喘和湿疹的病原。人接触螨虫的时间越长，就会变得越敏感。

Dust mites are only half a millimetre in size. Even though mites do not bite, sting or burrow, their droppings is the cause of allergies, asthma, and eczema. The longer we are exposed to mites, the more sensitive we become.

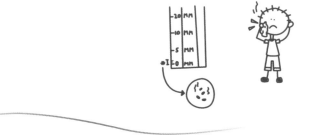

螨虫是蛛形纲的成员。1平方米地毯能住10万只螨虫。一个人一天会脱掉大约1克死皮。

Mites are members of the spider family. Nearly 100,000 mites can live in one square metre of carpet. A person loses about 1g of dead skin every day.

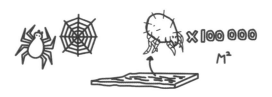

尘螨不喜欢冷而干的天气。温度低于20℃，湿度低于50%时，螨虫数量会减少。它们最大的敌人是阳光。

Dust mites do not like cold and dry weather. Temperatures under 20 °C and humidity below 50% lead to a decrease in the mite population. Their greatest enemy is sunlight.

雄性尘螨能活10—19日，而雌性尘螨最多能活70天。这使它有足够的时间产100个卵，排2 000粒粪便或颗粒物（从它身体里排出的小碎片）。

The male dust mite lives for 10-19 days, while a female will live for up to 70 days. This gives her enough time to produce 100 eggs, and 2,000 droppings or faecal particles (bits of waste expelled from her body).

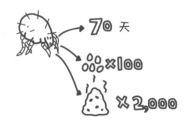

70 天

× 100

× 2,000

羊毛脂（羊毛蜡或羊毛干油）排斥尘螨，使我们有很好的理由用天然纤维替换合成纤维。

Lanolin (wool wax or wool grease) repels dust mites, giving us good reason to change from synthetic to natural fibres.

仅仅在美国，每年就有500万吨旧衣服和240万吨旧地毯被送进垃圾填埋场。一间200平方米的房子会产生4吨拆迁垃圾。

In the USA alone, 5 million tonnes of used clothing and 2.4 million tonnes of used carpets go to landfills every year. A 200 m² house creates 4 tonnes of demolition waste.

500万吨　　240万吨

How does it feel to realise that there are 100,000 tiny creatures sleeping with you in your bed?

当你意识到有10万个小生物与你一起在床上睡觉，你感觉如何？

你是愿意用化学药品杀死螨虫，还是愿意住在经过设计使螨虫无法生存的屋子里？

Would you prefer to kill the mites with chemicals or would prefer to live in a house that is designed in such a way that mites cannot survive there?

如果有人在房间里打喷嚏，你能想象打喷嚏是由几千颗螨虫粪便微粒导致的吗？

When someone in the room sneezes, can you imagine that the sneeze was caused by thousands of tiny mite droppings?

太阳只是提供阳光和温暖，使植物生长吗？还是它也会造成损害呢？

Does the sun only give light and warmth and make plants grow, or does it do any damage as well?

It is time to have a look   at the carpets in your house and at your school. Where are they placed? Are those rooms vacuum cleaned every day? And are they ventilated well? Are there many windows that can be opened and can the sun shine into the room and onto the floor? Or is there ultraviolet (UV) film on the windows to keep the sun from fading the carpets or the artwork on the walls? Once you have checked the room, take some dust scrapings from the carpet. Make sure you take dust from the deepest layer, where the nylon or woollen fibres are attached to the backing material. Place the dust under a microscope and watch the mites enjoy their meal.

看看你家里和学校里的地毯吧。地毯被放在哪里呢？那些房间每天都用真空吸尘器打扫吗？那里通风良好吗？有很多可以打开的窗户吗？阳光可以照到屋里、照到地板上吗？窗户上贴了防紫外线膜以防止地毯和艺术品褪色吗？等你完成了检查，取一些地毯上掉下的尘屑。确保在地毯的最深层（就是在尼龙或者羊毛织物紧贴背面材料的地方）取尘屑。把尘屑放在显微镜下，观察螨虫享受大餐的情景。

学科知识
Academic Knowledge

生物学	螨虫的生命周期从虫卵开始，随后变成有六条腿的幼虫、蛹，最后成为八条腿的成虫；螨虫与真菌之间存在共生关系：螨虫吃的多数食物是已经被真菌部分毁坏的东西；羊毛脂是产毛动物的皮脂腺产生的；羊毛脂有8 000到20 000种酯类。
化 学	生丝没有有害细菌和真菌；表层土在固碳和减缓气候变化中非常重要；从可再生资源中扩展蛋白质资源，以制造出多种多样的功能聚合物。
物 理	紫外线是一种消毒剂，因为它能摧毁核酸以杀死微生物或者降低它们的活性；紫外线是导致褪色的主要原因，因此办公楼的玻璃都会加上保护膜以保护地毯和艺术品不褪色，但是，使用保护膜会导致细菌、真菌和螨虫的增长。
工程学	在地毯制造中，用黄麻、剑麻、椰壳纤维和羊毛等天然纤维代替合成材料。
经济学	人们对紫外线保护膜制造产业和成本更高的空气过滤系统的需求不断上升；紫外线会使汽车油漆失去光泽，因此金属被加进车漆以保持光泽。
伦理学	不应只聚焦于单个问题的解决（如褪色），只重视表面效果，而不考虑可能给健康带来的副作用以及由此产生的额外成本。
历 史	1964年，尘螨被确认为主要的过敏原；古往今来，人们一直在寻找最明亮和最稳定的颜色，这使得胭脂红染料（从一种在墨西哥发现的甲虫身上提取的染料）在16世纪非常流行。
地 理	亚特兰大是美国佐治亚州的首府。
数 学	人们往往不考虑计算副作用（即外部化）所带来的资本成本和运转费用，如果把目前被社会承担的这些额外成本也列入原始模型一并考虑，投资回报将会减少。
生活方式	一些人盲目相信化学药剂是解决虫害问题的唯一方案。
社会学	人类对自然系统的无知以及对看不见的事物的漠视。
心理学	只有面对数据（如一个枕头中有10万只螨虫）和看到螨虫的照片，人们才会改变行为。
系统论	优先考虑健康而非外表的必要性，以及如何处理好二者之间的关系。

情感智慧
Emotional Intelligence

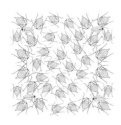

螨 虫

螨虫爱冒险，他们乐于尝试寻找新的居住地，对自己的好恶态度鲜明。螨虫擅长构建论据，讨论利弊得失。他们喜欢调查研究，不只看到了事物的表面，他们乐于亲自验证事实并获取第一手信息。他们有自知之明，知道体型非常小在某种程度上是他们自我保护的利器：如果没有人能够看到他们，就没有人能伤害他们。螨虫非常清楚他们对环境的影响。他们掌握了一系列事件的逻辑：人们不希望他们的地毯褪色，这给螨虫带来意外的保护，使他们免受太阳（螨虫最糟糕的敌人）照射。螨虫做了最坏的准备——忍受化学剂喷洒，但是他们知道作为一个家庭和一个物种，他们会继续生存下去，因为他们对化学剂已经具有耐药性，这使他们在这场斗争中占了上风。

艺术
The Arts

颜色对我们非常重要。紫外线会影响颜色，使其褪色。我们鉴定一下哪种颜色能在太阳每天都照射的情况下保持鲜亮和稳定。谁是创造不褪色的稳定颜色的大师？他使用了什么材料？找到更多可以制成颜料的东西，以及那些被持续用了成百上千年的颜料。现在与你的朋友一起组织一次绘画活动吧。

思维拓展
Systems: Making the Connections

建筑设计的重点是功能。由于制热制冷是主要的能源成本，建筑是隔热设计的，空气在建筑内循环使用，并通过过滤清除颗粒物和微生物。空调系统是用来保持恒温的，避免房间太冷或太热。吸音的地毯被用来改善对噪声和回声的控制，改善建筑物的整体氛围。特别是学校喜欢采用满铺地毯或拼接地毯来降低噪声。这些方法虽然有利于节能和降噪，却给居住者带来了患病态建筑综合征的风险。以照明营造良好氛围也已成为总体设施设计的一部分，而照明带来了额外的能源成本。建筑设计正朝着更加整体化的方向发展，将各种因素都考虑进来，包括健康。建造成本一直是考虑的重点，接下来是运转成本，最后是美学价值。很不幸，健康很少被优先考虑。促进进驻人员生产率的提高，往往是后期才想到的事，而不是建筑设计阶段的重点。最近的观察显示，很多从工程角度作出的决策，没有把居住者的健康考虑进来，例如，封闭窗户阻止新鲜空气流入。这一设计提高了能源效率。然而如果考虑到居住者的呼吸系统健康，这并非最优方案。此外，在窗户玻璃上贴上紫外线保护膜虽然有利于提高能源效率和保护房间内部设施，但是有很严重的健康隐患。这表明设计者缺乏对各种不同设计相互之间关联性的洞察力，而且把短期资本支出和年运行成本看得比居住者的长远健康更重要。人们不优先考虑健康，对决策如何影响健康也一无所知。我们需要一种更好的方案来同时达到多重目标，而无需在能源成本和健康之间寻找平衡。

动手能力
Capacity to Implement

做一个成本分析：是使用地毯成本低，还是使用无螨虫问题的其他地面铺装方式低呢？地毯需要每日吸尘，不能被紫外线照射，以免掉色。试着给生活质量估价：对你而言，低哮喘风险或低皮肤过敏风险的价值是多少？你应该有能力为你的选择、你对健康与外表的估价和你的结论而辩护，这是非常重要的。如果不同小组得出了不同的结论，与你的估价比较一下。然后，与你的父母和学校校长分享你的建议。

故事灵感来自
This Fable Is Inspired by

雷·安德森
Ray Anderson

　　雷·安德森（1934—2011）毕业于佐治亚理工学院，获工业工程学位。他在一家美国地毯龙头企业工作时，获得了地毯制造方面的第一手经验。1973年，雷创办了英特飞——一家专门制造地毯的公司。总部设在佐治亚州的英特飞公司从小做起，发展为地毯制造行业的领头羊之一，在4个国家设立工厂，生产的地毯销往100多个国家。雷提出了"零排放愿景"，承诺到2020年，该公司将通过产品和流程的再设计，消除对环境的所有负面影响，同时增加可再生材料和可再生能源的使用。雷著有《一个激进的实业家的商业经验》一书。

图书在版编目(CIP)数据

冈特生态童书.第三辑修订版:全36册:汉英对照 /
(比)冈特·鲍利著;(哥伦)凯瑟琳娜·巴赫绘;
何家振等译.—上海:上海远东出版社,2022
书名原文:Gunter's Fables
ISBN 978-7-5476-1850-9

Ⅰ.①冈… Ⅱ.①冈… ②凯… ③何… Ⅲ.①生态环
境–环境保护–儿童读物—汉、英 Ⅳ.①X171.1-49

中国版本图书馆CIP数据核字(2022)第163904号
著作权合同登记号图字09-2022-0637号

策　　划　张　蓉
责任编辑　程云琦
封面设计　魏　来李　廉

冈特生态童书
螨虫家园

[比]冈特·鲍利　著
[哥伦]凯瑟琳娜·巴赫　绘

何家振　译

记得要和身边的小朋友分享环保知识哦!
八喜冰淇淋祝你成为环保小使者!